A Day at the Beach

Ann and Paul Broadbent

This book belongs to

· ·

HODDER
EDUCATION
AN HACHETTE UK COMPANY

It is a hot, sunny morning. Max and Bel are going to the beach with Dad.

I've got a big bucket.

Find the opposite pairs.

Point to an open car door. What is opposite to open? What is opposite to big? Point to a big spade. Point to a small bucket. Point to a closed parasol.

Mum and Jasmin will join them later.

Talk about the position of objects.

What is under the car? Where is Max? Where is Dad's hat?
What is next to the parasol?

At the beach, Bel and Max race to a rock pool. They count the fish and the sea creatures.

I can see 8 fish altogether.

Find the totals.

How many big crabs? How many small crabs? How many crabs in total?
Count the thin fish. How many round fish? How many fish in total?

Look! I've caught 2 fish!

They are so busy that they don't notice Mum arrive with Jasmin.

Try subtracting and find how many are left.

8 fish take away 2 fish. How many fish are left?
Cover some shells. How many shells are left?

It is time to eat lunch and drink some cool orange juice.

Compare the size and length of objects.

Find the small glass. Point to the bananas in order; start with the longest banana. Point to the biggest tomato. Point to the short straw.

Bel turns her back for a moment and her glass is empty!

Talk about full and empty.
Which glasses are full? Which glass is empty? Which bucket is full?

Max and Bel make four fantastic sandcastles.

It wasn't me. Who could it be?

Find flat shapes and solid shapes.

How many circles can you see? Which objects are the same shape as the bucket? Point to a triangle. How many squares can you see?

These footprints look a bit small.

They turn their backs for a moment and one sandcastle has been knocked down!

Talk about the lines and patterns.

Follow the zigzag and wavy lines with your finger. Describe the colour pattern on the ball. Describe the patterns on the sandcastles.

Max and Bel play mini-golf with Dad.

Crazy Golf

Who took my ball?

Read the numbers to 10.

Read the numbers on the flags. Count the number of spots on each flag. Which flags have the wrong number of spots?

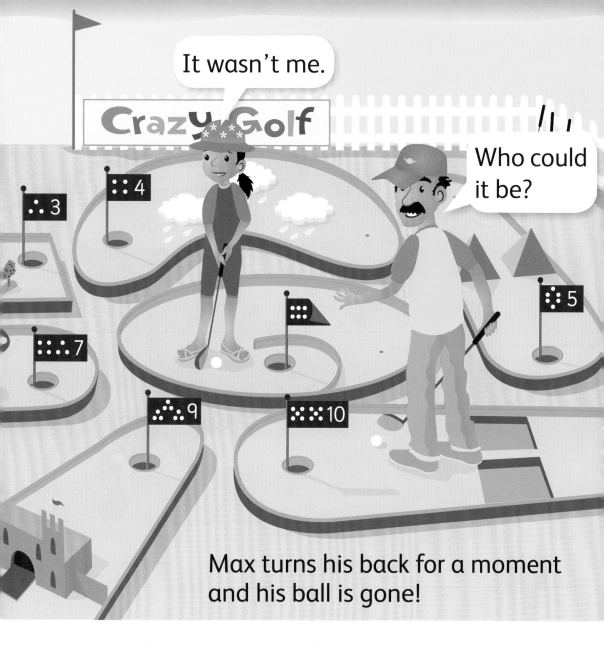

Max turns his back for a moment and his ball is gone!

Count the numbers on the flags in order.

Which number comes after 3? Which number is before 8?
Which number is missing?

Max and Bel are tired after having fun on the beach all day.

Let's have a rest in the shade.

Talk about times in a day.

Where is Jasmin in the morning? Where is she at lunchtime?
Where is she in the afternoon when Max and Bel are resting in the shade?

What a surprise!
Max and Bel were so busy playing that they
didn't notice Jasmin at the beach.

Talk about the sequence of events in this story.
*Which object did Jasmin collect first – the shell, the golf ball
or the straw?*

Write the missing numbers on the flags.

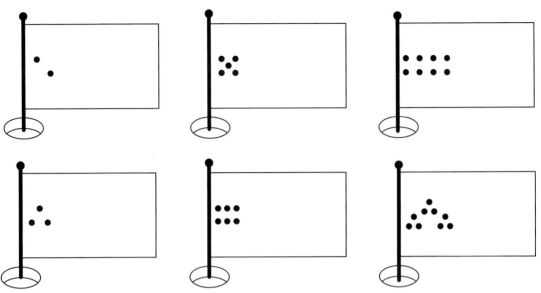

 Join the groups of balls to the correct numbers.

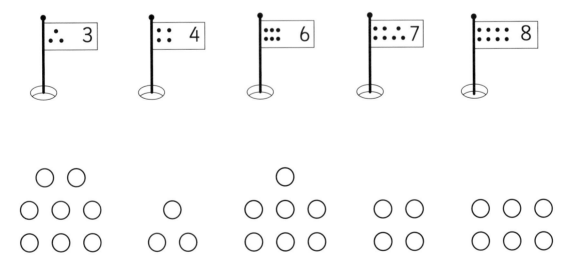

Join the opposite pairs.

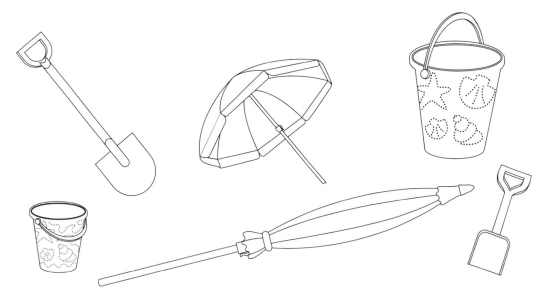

 Colour these to make patterns.

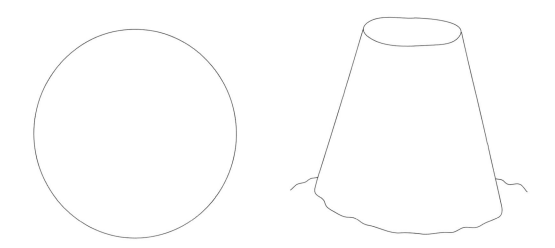

Count each group. Write the number.

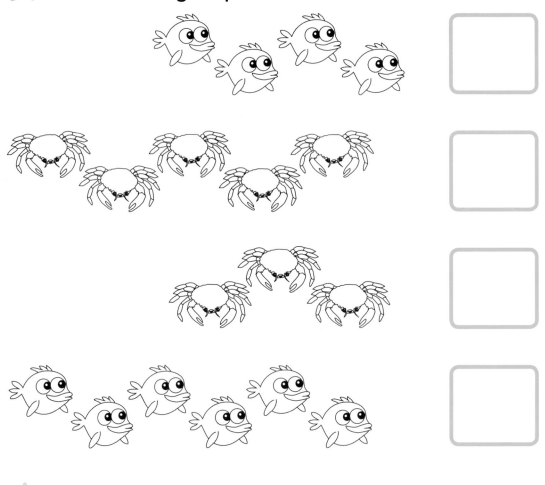

 Count the fish and crabs above.
How many altogether?

 crabs
altogether

 fish
altogether